Parry of the Arctic

Books for Younger Readers by Pierre Berton

The Golden Trail
The Secret World of Og

ADVENTURES IN CANADIAN HISTORY

The Capture of Detroit
The Death of Isaac Brock
Revenge of the Tribes
Canada Under Siege

Bonanza Gold
The Klondike Stampede

Parry of the Arctic
Jane Franklin's Obsession

PIERRE BERTON

PARRY OF THE ARCTIC

ILLUSTRATIONS BY PAUL MC CUSKER

An M&S Paperback Original from
McClelland & Stewart Inc.
The Canadian Publishers

An M&S Paperback Original from McClelland & Stewart Inc.

First printing January 1992

Copyright © 1992 by Pierre Berton Enterprises Ltd.

All rights reserved. The use of any part of this publication reproduced, transmitted in any form or by any means, electronic, mechanical, photocopying, recording, or otherwise, or stored in a retrieval system, without the prior written consent of the publisher – or, in case of photocopying or other reprographic copying, a licence from Canadian Reprography Collective – is an infringement of the copyright law.

Canadian Cataloguing in Publication Data

Berton, Pierre, 1920-
Parry of the Arctic

(Adventures in Canadian history. Exploring the frozen world)
"An M&S paperback original."
Includes index.
ISBN 0-7710-1434-1

1. Parry, William Edward, Sir, 1790-1855 – Juvenile literature. 2. Arctic regions – Discovery and exploration – Juvenile literature. 3. Northwest passage – Juvenile literature. 4. Inuit – Juvenile literature.* 5. Explorers – England – Biography – Juvenile literature. I. McCusker, Paul. II. Title. III. Series: Berton, Pierre, 1920- . Adventures in Canadian history. Exploring the frozen world.

G635.P3B4 1992 j910′.9163′2 C91-095530-1

Cover design by Tania Craan
Text design by Martin Gould
Cover illustration by Scott Cameron
Interior illustrations by Paul McCusker
Illustration "First communication with the natives of Prince Regent's Bay" by John Sacheuse, p. 15, courtesy The Metropolitan Toronto Library
Maps by Geoffrey Matthews
Editor: Peter Carver

Typesetting by M&S

Printed and bound in Canada

McClelland & Stewart Inc.
The Canadian Publishers
481 University Avenue
Toronto, Ontario
M5G 2E9

Contents

Parry of the Arctic

ARCTIC OCEAN
Edge of polar ice
BEAUFORT SEA
ALASKA
PRINCE PATRICK ISLAND
MELVILLE ISLAND
M'Clure Strait
BANKS ISLAND
Viscount Melville Sound
VICTORIA ISLAND
Arctic Circle
Great Bear Lake
Coppermine
Coronation Gulf
Queen Maud Gulf
Mackenzie River
Coppermine River
Great Fish River
C A N A D A
Great Slave Lake
Lake Athabasca

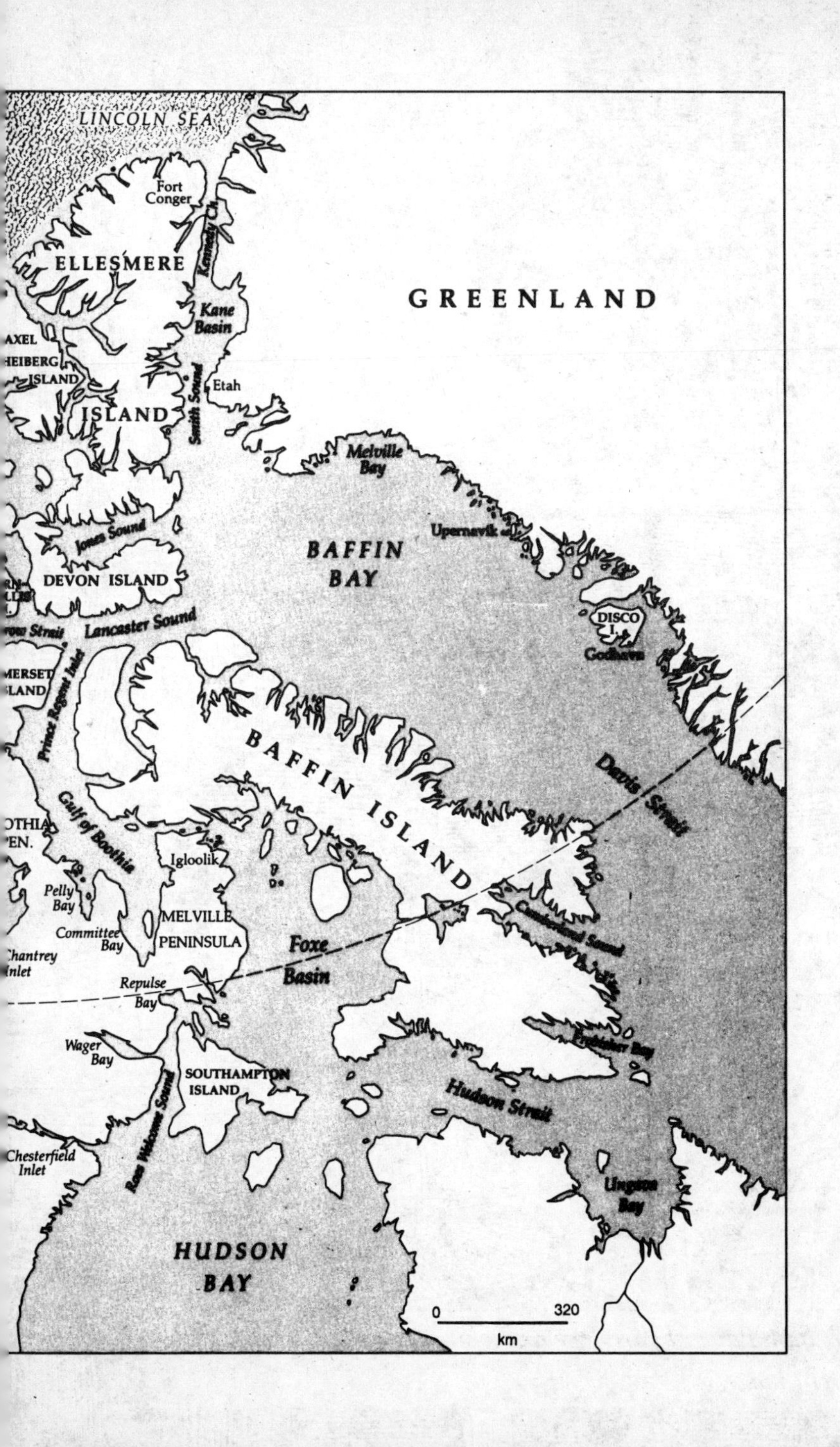
LINCOLN SEA
Fort Conger
Kennedy Ch.
ELLESMERE
ISLAND
Kane Basin
Smith Sound
Etah
GREENLAND
Melville Bay
Upernavik
Jones Sound
DEVON ISLAND
BAFFIN BAY
Lancaster Sound
DISCO I.
Godhavn
Prince Regent Inlet
BAFFIN ISLAND
Davis Strait
Gulf of Boothia
Igloolik
Pelly Bay
Committee Bay
MELVILLE PENINSULA
Foxe Basin
Cumberland Sound
Repulse Bay
Wager Bay
SOUTHAMPTON ISLAND
Roes Welcome Sound
Hudson Strait
Chesterfield Inlet
HUDSON BAY
0
320
km

William Edward Parry, the first white explorer to penetrate the Arctic maze.

Chapter One

The strange people

When Edward Parry was just thirteen years old, he left grammar school in England and joined the Royal Navy as a young midshipman. That was the ambition of many young English boys – an ambition that could be fulfilled if you knew the right people and were bright enough to be accepted into the service. Midshipmen were naval apprentices. Most who joined at the age of thirteen would rise to become officers in the Navy, and many would become admirals. Parry became one of the greatest explorers of his era.

That was, of course, the ambition of many young men in the nineteenth century. Explorers were the great heroes of the era – an era that has been called "The Golden Age of Exploration." To unravel the secret of the source of the Nile, to enter into darkest Africa, to explore the unknown islands of the South Seas, or to face the mists of the Arctic in a search for the fabulous North West Passage – this was the goal that they sought.

Explorers were as wildly popular then as movie stars or

rock stars or television stars are today. They were knighted by the queen. They wrote best-selling books and articles. They were given huge sums of money as prizes. They were pursued in the streets by fans. They were courted by women who wanted them at their society tables. Their lectures were attended by thousands. In short, they were as famous as the prime minister himself.

Parry was just twenty-eight years old when the Royal Navy put him in charge of a ship – one of two that would try to search out the North West Passage. In those days the Canadian Arctic was an unknown quantity.

No one knew what existed at the top of the North American continent. One man, Alexander Mackenzie, had actually reached the Arctic Ocean in 1789 when he came down the river that bears his name to its mouth. Another, Samuel Hearne, had reached the mouth of the Coppermine in 1771. Apart from these two pinpoints on the map, nobody knew anything about the Arctic coast that was supposed to form the northern roof of the continent – indeed, nobody was quite sure whether there *was* an Arctic coast.

Hudson Bay had been partially explored and men had reached the tip of Baffin Island. But what lay beyond and to the west? Was it open ocean? Was it solid land? Was it a whole tangle of islands and channels? And was there a Passage that would link the two great oceans, the Atlantic with the Pacific?

The wars with Napoleon ended in 1815. Because the Navy had nothing for its young men to do, it set about to explore the oceans of the world – including the Arctic. Three years after the end of the Napoleonic wars, the Navy sent two ships into that frozen Arctic world. The lead ship was the *Isabella,* captained by an old naval hand named John Ross. The smaller ship, the *Alexander,* was placed under the charge of Edward Parry.

In mid-June, 1818, the expedition had crossed the Atlantic and entered Davis Strait between Baffin Island and Greenland. The officers and crews now had their first view of the ice-bound sea in all its splendour and all its menace. Here was a brilliant world of blue, emerald and white – dazzling to the eye. To some, the great frozen mountains – icebergs – that whirled past seemed to have been carved by some mysterious artist. Here were cathedrals and palaces, statues and castles, all brilliant white, flashing in the sun's rays, each slightly blurred as in a dream – a world of shimmering ice. Colours were intense. As Parry's superior, Ross, wrote: "They glitter with a vividness of colour beyond the power of art to represent."

Both men were awed by the strangeness of the savage realm they had entered. Soon they would be in unknown waters. But it was comforting to encounter the fleet of three dozen whaling ships, all flying the British flag, and to hear the cheers of the whalers as they passed through. This was where civilization ended.

The whalers gave Parry and Ross the first warning about the difficulties of the Arctic climate, which changed from year to year. This year the ice was much worse than expected. The previous winter had been the worst in ten years. The whalers could hardly find a clear passage south through the ocean of icebergs.

On July 1, the two ships entered a frightening maze of ice. Parry tried to count the icebergs and gave up when he reached a thousand. For the next month the two ships worked their way north along the Greenland coast, blinded by fog and almost crushed by the pressure of the ice-pack, that vast floating ocean of solid ice, during one terrible gale. It was a close call: the sterns of the two vessels bashed into each other. Spars, rigging, life boats were torn apart. But they survived.

A day or so later, at the very western tip of Greenland, they came face-to-face with an unknown culture. These were the Inuit, whom they then called Eskimo. Even John Sacheuse, their native interpreter from South Greenland, had never heard of this strange race of polar people, whom Ross named "Arctic Highlanders." Sacheuse could understand their dialect only with great difficulty.

Here is a picture of this first encounter between white men and Inuit people. Notice that the Inuit are dressed as one might expect, in clothes made from fur and sealskin – perfect for the Arctic climate. But the two men who greet them – Ross and Parry – are dressed exactly as they would have been if they'd gone to the South Seas. They wore their cocked hats, tail coats, white gloves, and swords, as if they were attending a party in London. The fact was, of course, that no exploring nation had yet understood the need for special clothing in the North.

If the British officers were baffled by these new squat creatures muffled in furs, the natives were equally baffled. "Where do you come from, the sun or the moon?" they asked. And so the picture was made, not by the British, but by Sacheuse himself. He was a young Christianized native from the South who had stowed away two years before on a sailing ship and eventually reached England where he studied drawing. He was the first Arctic native encountered by British explorers.

A very strange and almost comic scene followed. The Inuit hung back, terrified of the strange people on the ships. And so it was decided that one of Parry's officers would go out with a white flag on which was painted the white man's emblem of peace – a hand holding an olive branch.

The only trouble was, the natives didn't know what an olive branch was or what it was supposed to mean. No olive trees grew on that bleak shore. In fact, no trees grew at all. Yet none of the white men seemed to think this strange. Ross, however, was more practical. He put a flag on a pole and tied a bag full of presents to it and that worked very well.

These strange people had had no contact with the world beyond their own region. They were surprised at the presence of Sacheuse. It hadn't occurred to them that there might be others like themselves in the world. As for the men with the white skins, they thought they had come from the sky.

These were landsmen. They didn't know anything about

boats. They had never even seen a boat. Even the native word "kayak" had no meaning for them. They spoke to the ships as if they were living things. "We have seen them move their wings," they said. Sacheuse tried to explain the ships were floating houses. They had trouble believing him.

They were baffled by their first glimpse into a mirror and tried to discover the monster they believed was hiding behind it. They laughed at the metal frames of the eye-glasses worn by some of the sailors. When they were offered a biscuit, they spat it out in disgust. They wondered what kind of ice the window panes were made of. They tried to figure what kind of animals produced the strange "skins" the officers were wearing. When they were shown a watch, they thought it was alive and asked if it was good to eat. The sight of a little pig terrified them. A demonstration of hammer and nails charmed them. The ship's furniture baffled them, for the only wood they'd ever seen came from a tiny shrub, whose stem was no thicker than a finger.

Sacheuse made them take off their caps in the presence of the officers. That suggests how quickly he had absorbed the white way of life. It was the first attempt – one of many that would be made over the coming years – to "civilize" the natives.

They obeyed cheerfully enough. But they must have been as mystified by this ritual as the English were to find that human beings actually lived in this strange and cruel land. Yet nobody asked how it was that a band of people, who couldn't count past ten, had managed to adapt to the savage wilderness of Greenland. If they had done so, it would have been easier in the future for British sailors to adapt to the Arctic way of life.

The expedition moved on to the very top of Baffin Bay and then sailed west to the southern tip of what is now Ellesmere Island. It turned south still seeking a channel that might lead to the North West Passage.

Then, at the end of August, they encountered a long inlet leading westward, which an earlier explorer had named for Sir James Lancaster. Was this the way to the Orient? Or was it simply a dead end? Nobody knew.

Parry was full of optimism. He was sure this was the route that would lead, if not to the Russian coast, at least into the heart of the Arctic to connect with other lanes of water to the west.

Ross wasn't so sure. As the two ships moved into the unknown, he became convinced that no passage existed. Then, one foggy afternoon at the end of August they reached the thirty-mile (48 km) point, and Ross pulled up waiting for Parry to catch him and for the weather to clear. The officer of the watch roused him in his cabin to announce the fog was lifting. And this is where Ross made the blunder that wrecked his career and paved the way for Edward Parry to achieve greatness.

Ross saw, or thought he saw, a chain of mountains blocking all access to the west. Nobody else saw it. Was it an Arctic mirage that baffled Ross? These would become well known in the years that followed. Or was Ross seeking an excuse to end the journey, as many believed? To the stunned surprise and anger of the others, Ross, without a word of explanation, turned about and headed for home "as if some mischief was behind him."

As William Hooper, the purser of Parry's ship, wrote, "Thus vanished our golden dreams, our brilliant hopes, our high expectations!" Parry and his crew were bitterly disappointed. They had expected to spend a year in the Arctic, and now here was Ross high-tailing it back for England.

The Navy was not amused. They sided with Parry, retired Ross on half pay, and never used him again. The following year they sent Parry off at the head of an expedition to find the North West Passage.

Actually, Parry was no better qualified for Arctic exploration than the unhappy John Ross. Parry owed his position to John Barrow, Jr., the moon-faced bureaucrat who was the unseen power behind the British Navy. Barrow, the civil servant – he was actually *second* secretary to the Admiralty, though he acted as if he were the first – has been called the father of modern Arctic exploration. It was he who sent ship after ship around the world charting unknown waters off the coasts of Africa and Asia. But it was the Arctic that obsessed him.

He held the curious belief that somewhere beyond the impenetrable ice, there lay a warmer ocean, surrounding the North Pole but walled off from the world by a frozen barrier. He was dead wrong – how could you have warm water surrounded by a doughnut-shaped circle of ice? Yet

Barrow's belief in an "Open Polar Sea," as he called it, persisted for half a century.

One man who knew that was nonsense was a whaling captain, William Scoresby, the most courageous and skillful skipper in the Greenland fleet. He was more than that; a student of philosophy and science, and an inventor, he was the leading expert on Arctic conditions, and was about to complete a monumental work on the subject.

He would have been a better choice than Parry to lead the first exploration into the Arctic puzzle. But it was Parry, the regular naval man, who got the nod from the powerful

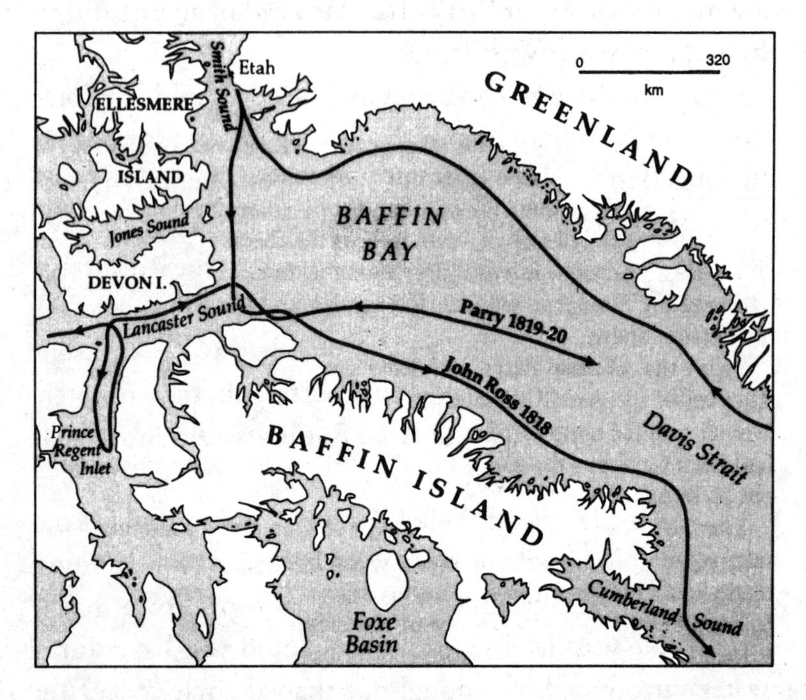

The start of the quest

John Barrow. Barrow was positively chilly, if not rude, when Scoresby approached him, suggesting he be allowed on the expedition.

In its arrogance the Navy scorned whalers. Under the rigid English class system, they weren't "gentlemen." Neither was Barrow, who came from humble stock, or even Parry, though his father, a doctor, had members of nobility as patients.

But the Navy was convinced its own officers and men were capable of anything. Had Nelson not defeated the French at Trafalgar? Besides, Parry had powerful friends and, in the Royal Navy of 1818, it was who you knew, not what you knew, that counted.

Chapter Two

Into the unknown

Edward Parry belonged to a new generation of explorers. He was the model by which those who followed would be judged. He believed that an Englishman could overcome any obstacle. Devout, steadfast, and loyal, he believed in hard work and team spirit.

A handsome officer, tall, slightly stooped, with curly chestnut hair and soft grey eyes, he was well-spoken and eager to please. He was not an excitable man. His journals did not make too much of the hardships he faced. But he was lucky. He sailed north at the right moment, when the Arctic channels were clearer of ice than they had been in a decade.

But his real achievement lay not in opening a passage to the Arctic islands. His greatest accomplishment was his understanding of his crew and his determination to keep them healthy in mind as well as body.

These were tiny ships by our standards. You could set both of them down, end to end, in a modern football field and still have some room left over. Being ships of war, they were made for the open sea, not the shallow coastal waters of the Arctic maze. Crammed together in these close quarters, sleeping in hammocks, buffeted by mountainous waves and shrieking gales, unable to leave their vessels during the long Arctic night, each seaman's good nature was tested and often found wanting. The food was dreadful – mainly salt pork – while the beer, which was their main beverage, grew stale and sour after months in barrels. Their confinement was as prison-like as that suffered by modern astronauts, but unlike these twentieth-century explorers, they had no support system, no information network, no form of electronic monitoring. They were on their own – out of touch with the world for months and usually for years. Yet they survived, and this was Parry's achievement.

Parry realized that the greatest peril of wintering in the Arctic would not be the cold. It would be boredom. For as long as ten months nothing moved. The ships became prisons. The masts and superstructure had to be taken down. The hatches were hermetically sealed. The ships were smothered in blankets of insulating snow.

Given the cramped conditions, the best disciplined seaman could break down. Small irritations could be magnified into raging quarrels. Imagined insults could lead to mutinous talk and even mutiny. Parry was determined to cope with the monotony of the Arctic winter. All the officers under him were young men – for the Arctic required youth, energy and physical fitness. Parry and his crew had these qualities.

By April 1819, the two ships in their fresh coating of black and yellow paint were ready to set off. Down at the docks at Greenwich the well-wishers flocked. Parry wrote that "no other expedition had ever attracted a more hearty feeling of national interest." One of the visitors particularly attracted him – a Miss Browne who was the niece of one of his officers. They flirted together. Perhaps Parry thought that when he returned he would marry her.

The ships left on May 11 heading across the Atlantic. Parry's instruction was to head directly up Davis Strait to Lancaster Sound – the same opening in the wall of mountains that had fooled John Ross. If it turned out that there *were* mountains, as Ross had thought, Parry was ordered to go on north and find another entry into the Arctic world.

His task was to get right through as quickly as possible, deliver his documents to the Russian governor at Kam-

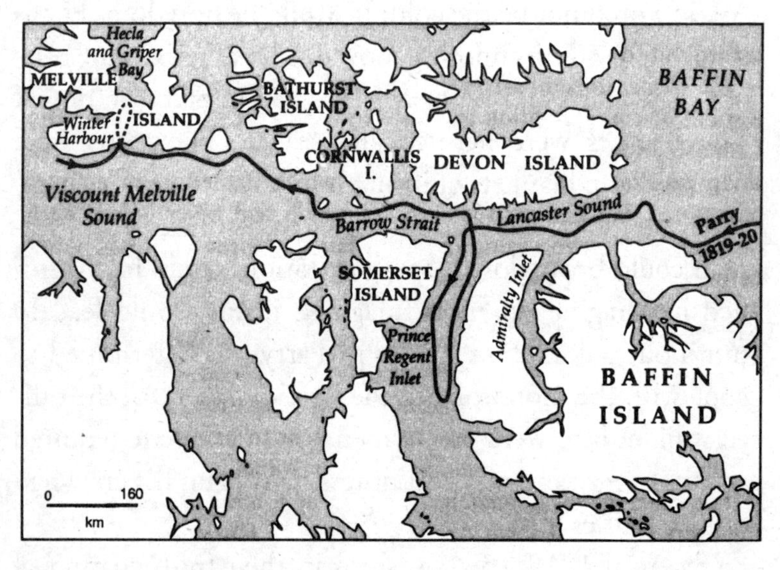

Parry's first voyage, 1819-20

chatka, and then sail on to Hawaii and return home. His superiors were over-optimistic about the Arctic. They believed Parry could get through perhaps in one year and certainly in two. But Parry thought that the possibility of getting into China that year was "too much to hope for." Too much indeed! Almost ninety years would pass before anybody made it.

Parry's chances of bulling his way through were slimmer than he knew. In a bad season the odds were about one in a hundred, in a light season, about fifty-fifty. Even in an exceptional season – and this *was* an exceptional season – there was still a 25 percent chance of failure.

Parry's first setback came when he ran into the great river of ice that breaks off from the polar pack and pours down the centre of Davis Strait. The ice-pack was impenetrable – a vast ocean of icebergs all frozen together. Dozens of feet deep, it stretched off to the horizon as far as the eye could see – a great, rumpled sheet of white that even today can resist a modern icebreaker. Nonetheless Parry decided boldly to try to force his way directly through. That shortcut would save weeks. He would use his larger ship, *Hecla,* as a battering ram to clear the way for the weaker ship, *Griper.*

That didn't work. Unable to get through the barrier he turned north for three weeks and crossed the Arctic Circle into Baffin Bay. There he tried again to bull his way through by brute force.

His crews experienced for the first time the exhausting

toil that would be a feature of Arctic exploration for all of the nineteenth century. They had to sit in small boats, straining at the oars, as they attempted to tow the big vessels through the ice-choked channels. They attached cables to anchors in the icebergs and pulled the ships westward, foot by foot. They trudged along the ice floes, clinging to ropes like tug-of-war teams, hauling the vessels in the direction of their goal.

It was disappointing work. On one long day they sweated for eleven hours and moved no more than four miles (6.4 km). Once, when the *Hecla* was trapped, they worked for seven hours with ice saws to cut her free, only to find her frozen in again at the end of the day. Parry urged them on with extra rations of rum and meat. In the distance they could hear the music of the great black whales, a shrill ringing sound, rather like hundreds of musical glasses badly played.

At last they were through. The broad entrance to the Lancaster Sound lay directly ahead. The towering mountains of Bylot Island crowned its southern entrance. Now Parry knew that the next few days would either make his reputation or break it. He couldn't wait for the slower ship, *Griper.* On August 1, as soon as the wind was favourable, he set out himself up the sound under full sail.

The weather was clear. The mysterious channel lay open. As the wind increased to a gale, Parry could sense that everyone on the ship was holding his breath. Nobody had sailed up this lane of water since the days of the Iceland

The crow's nest was a barrel lashed to the main mast, where a lookout could spot new channels and icebergs.

fishermen six centuries before. Everything was new. Long inlets – or were they only bays? – led into the unknown from both shores. Broken hills in the south rose one above the other to snow-clad peaks. Was there land ahead? Was the route to the Passage blocked as Ross had believed? No one knew.

But the lookout, high up in the crow's nest – a barrel tied to the top mast – could find no hint of any mountains. The sound – eighty miles (129 km) wide – appeared to be clear of ice. For two days the ship moved on. Now all hands began to feel a sense of relief. There were some who began to believe the Passage could be mastered.

At six o'clock on the evening of August 4, their hopes were dashed. The lookout reported land ahead and everybody's spirits fell. But then, the land turned out to be a small island, and it was now obvious that the channel ahead was clear. John Ross's failing eyesight had played him false. Parry, who had never believed in these mysterious mountains, felt vindicated.

Parry's smaller ship, the *Griper,* joined him here. Parry's purser, William Hooper, wrote that "there was something peculiarly animating in the joy which lighted every countenance.... we've arrived in a sea which had never before been navigated, we were gazing on land that European eyes had never before beheld ... and before us was the prospect of realizing all our wishes and exalting the honour of our country. ..."

Parry could now see that a water highway stretched straight as an arrow directly into the heart of the unknown

Arctic islands. But he was groping with the unknown. He was like a man travelling through a long tunnel, able only to guess at the mysteries that lay to the north and south.

He could see precipices on the north shore, cut by chasms and fjords, some of them rising for five hundred feet (152 m). On the south he could see a tableland, cut by broad channels, one of them more than forty miles (64 km) wide. But was this the route to the Bering Sea and the Pacific Ocean? He had no way of knowing.

Blocked by ice ahead of him, he turned south into a new channel forty miles (64 km) wide which he named Prince Regent Inlet. The two ships sailed down it for more than a hundred miles (160 km) past the snow-choked ravines and high rock walls of a great island (Somerset). There, more ice barred his way.

Now he faced a new dilemma. Perhaps this was only a bay after all! And so he turned about and sailed back to Lancaster Sound. He found some open water along the north shore but was held up again by snow and sleet.

Finally the weather cleared and he headed west into the very heart of the mysterious Arctic.

He left the high cliffs and peaks above the coast line of Devon Island and plunged into the northern mists of a broad channel, which he named for the Duke of Wellington. He swept on, past the rock walls of Cornwallis Island and the wriggling fjords of Bathurst Island. Finally, he entered an immense inland sea that he named Viscount Melville Sound after the first lord of the Admiralty.

Parry and his crew had been promised a five-thousand-pound reward by Parliament if they could cross the meridian at 110°. On Sunday, September 4, they did it. For his feat Parry would be given a thousand pounds – a small fortune at that time, worth at least $100,000 in 1991.

It was well earned. In one remarkable five-week sweep he had explored some eight hundred miles (1,287 km) of new coastline. And there was more: to the north he could see the twelve-hundred-foot (366 m) cliffs and rugged highlands of another great island, which he also named for Viscount Melville.

Eighteen days later, with the weather getting worse and more ice forming, he gave up the struggle. This was as far as he could go that year. He would have to go into winter quarters in a small bay on Melville Island's south shore. That would be his prison for the best part of a year.

Now the back-breaking work began. The crew laboured for nineteen hours without a break in the ghostly glimmer of the northern lights, sawing a channel, square-by-square, in the bay ice. Every time they cut it open it seemed to refreeze before their eyes. But after three days they had managed to cut an opening two and a third miles (3.7 km) long into the bay. There the ships would rest for more than eight months protected from the fury of the sea by a reef of rocks. Parry named it Winter Harbour.

Now these Englishmen were marooned at the very heart of the darkest and most desolate realm in all the northern hemisphere. The nearest permanent civilized community lay twelve hundred miles (1,930 km) to the east on an island off Greenland's west coast. The nearest white men were the fur traders on Great Slave Lake, seven hundred miles (1,127 km) to the south. Another twelve hundred miles to the southwest were the uninhabited shores of Russian Alaska, and beyond that, Siberia. To the north, a frozen world stretched, stark and empty, to the Pole.

Thus for hundreds of miles in every direction the land was devoid of human life. The nearest Inuit were close to five hundred miles (800 km) away. Soon the wildlife itself would vanish. New species, which they had seen – musk oxen, caribou, and lemmings – would be gone, and so would the gulls and the terns. The only light they would have would be the flickering candles, rationed carefully to one inch (2.5 cm) a day. For the sun itself would depart with the animals.

These were the first white men to winter in the Arctic. Under such conditions Parry knew that men could become half-crazed. His purpose was to keep his crew so busy that they wouldn't have time to brood. There would be plenty of daily exercise, regular inspections, and afternoons crammed with work, much of which Parry would invent. The emphasis was on physical health, cleanliness, and "busyness."

At 5:45 in the morning the men were up scrubbing the decks with warm sand. They had breakfast at eight, were inspected right down to their fingernails at 9:15, and then set about running around the deck or, in good weather, on the shore. They were kept occupied all afternoon, drawing, knitting yarn, or making gaskets. After supper they were allowed to play games or sing and dance until bedtime at nine. The officers spent a quieter evening. They read books, wrote letters, and played chess or musical instruments.

Around them in the gathering gloom, the land stretched off, desolate and dreary, death-like in its stillness, offering no interest for the eye or amusement for the mind. Parry noted that if he spotted a stone of more than usual size on one of the short walks he took from the ship, his eyes were drawn to it and he found himself pulled in its direction. So deceptive was the unchanging surface of the snow that objects apparently half a mile (800 m) away could be reached after a minute's walk.

In such a landscape it was easy for a man to get lost. As a result, Parry forbade anyone to wander from the ships. And when darkness fell, their isolation was complete. In mid-season one could, with great difficulty, read a newspaper by daylight – but only at noon.

He had no idea how cold it could be in the Arctic. He found that the slightest touch of a bare hand on a metal object tore off the skin. A telescope placed against his eye burned like a red-hot brand. Leather boots were totally impractical because they froze hard and brought on frost-bite. Parry devised a more flexible footwear of canvas and green hide.

Sores wouldn't heal. Lemon juice and vinegar froze solid and broke their containers. The very mercury froze in the thermometers. When doors were opened, a thick fog poured down the hatchways, condensing on the walls and turning to ice.

Damp bedding froze, forcing the men into hammocks. Even the steam rising from the bake ovens froze – which meant that there was less bread to eat. And there wasn't enough fuel to heat the ships. The crews were always cold. The officers played chess bundled up in scarves and great-coats.

Nevertheless, the expedition produced and printed a weekly newspaper to which Parry himself contributed. They put on plays and skits, the female impersonators shivering gamely in their thin garments. Parry admitted it was almost too cold for the actors and the audience to enjoy the shows. No wonder! In his own cabin the temperature dropped in February to just seven degrees Fahrenheit (–14°C).

By mid-March, twenty men were sick and Parry began to look ahead. When would the thaw come? How long must they remain in prison? It was still bitingly cold a month later. Parry hadn't figured on that. He now began to have some doubts about getting through to the west.

Two weeks dragged by. The sun now shone at midnight. The temperature moved back to the freezing point. Game began to appear – a few ptarmigan and, a month later, caribou. The fresh meat reduced the danger of scurvy, although one man died from it at the end of June.

Parry decided on a two-week trip across the big island. He took four officers and eight men. They dragged eight hundred pounds (363 kg) of equipment on a two-wheeled cart. Strangely, it didn't occur to them to use dogs and sledges as the Inuit did.

Then July arrived. That was the only bearable month on the island. Yet ice still choked the harbour. Parry was desperate to be off. His sails were ready for an immediate start. He realized how little time he had – nine weeks at the most – a painful truth he couldn't hide from the crew.

The days that followed in late July and early August were maddening. The ice melted. They moved forward. The ice

blocked their way. They anchored. The ice shifted. They moved again. The wind changed. The ice moved back.

On August 4, they were able at last to set off into the west, but again the ice stopped them. The floes closed in on the *Griper* hoisting her two feet (0.6 m) out of the water.

Parry sent an officer ashore to climb a high cliff and look over the frozen sea to the west. He reported land fifty miles (80 km) away, but the sea itself was covered with ice floes as far as the eye could reach. They were so closely joined that no gleam of water shone through. Parry named the new

Parry's two-wheeled cart was a cumbersome vehicle for moving equipment over the Arctic wastes.

land Banks Island after the president of the Royal Society, but he did not reach it.

Now his optimism began to fade. The previous summer the Passage had seemed within his grasp. All winter he had planned to break out of the harbour and to sail on to the Bering Sea, but now the Arctic was showing its real face.

The ice kept him in prison for five more days. When it cleared he tried to go northwest. Again the ice stopped him. He ran east looking for a southern way out, but once more found himself frozen in. Like a rat in a trap, he was scurrying this way and that, trying to escape.

On August 23, he managed to reach Cape Providence on the southern tip of Melville Island after performing "six miles (9.6 km) of the most difficult navigation I have ever known among ice." He couldn't know then that he was facing the dreaded ice stream that flows down from the Beaufort Sea where the ice is fifty feet (15.2 m) thick. The polar pack squeezing down past Banks Island into Melville Sound and on to the channels that lead south and east to the North American coast is all but impassable. One hundred and twenty-four years would pass before the motor of the tough little RCMP schooner, *St. Roch,* would finally push it through the barrier on the eastern side of the present Banks Island.

Parry was heartsick. Now he had to admit that he couldn't do it. It wouldn't have given him much comfort if he had known the truth – that no sailing vessel would ever conquer the Passage, and no other vessel either, in his century. He had gone as far as any man could go in the primitive conditions of that time.

He had a decision to make. He knew he could stretch his food and fuel for another winter by careful rationing. But he couldn't answer for his crews' health. And so he turned his ships east, hoping to find an alternate Passage to the south. None appeared. At the end of August he set off for England and was home by the end of October, with all but one of the ninety-four men who had gone north with him.

He'd been deceived by the vagaries of Arctic weather. He hadn't reckoned on the severity of the climate, or the shortness of the season. He was convinced, quite rightly, that his chosen route was impossible. If the Passage was to be conquered, another way must be found.

Still, he remained an optimist. Promoted to commander, cheered up by the applause of the politicians, the congratulations of the Navy, and the cheers of the public, he could be pardoned for believing that the next time he would make it. But almost nine decades would pass before any white explorer travelled from the Atlantic to the Pacific by way of the cold Arctic seas.

Chapter Three

The ice won't budge

PARRY BECAME the nineteenth century's first hero explorer. These were folk figures larger than life. Their failings, flaws, and human weaknesses were ignored by the public and the press, who saw them as leaders in a growing British Empire.

Fortune accompanied fame. Parry had his thousand pounds from Parliament – an enormous sum in those days. Now he got another thousand from a British publisher for the rights to his journal. Letters of congratulations poured in. He might not have discovered the Passage, but he had, in his phrase, "made a large hole in it."

His time was occupied by a round of social events that might have turned the head of a more excitable officer. His portrait was painted by a member of the Royal Academy. He was given the freedom of his native city, Bath. He was presented at court. The new king, George IV, offered congratulations. London hostesses scrambled to invite him to dinner. Exclusive clubs asked him to become a member. The first of the Arctic heroes was setting a pattern that others would seek to copy. Considering the rewards, who would not dare to brave the Arctic blasts?

And certainly Parry was eager to be off again. That winter of 1820-21, he began to prepare for his second voyage to conquer the Passage. Indeed, Parry felt he was in a race. His main fear was that the Russians would beat the English to it.

Before the year was out, the decision was made. Again there would be two ships, the *Hecla,* and a new ship, the *Fury,* which was the *Hecla*'s sister. Parry would command the *Fury.* The *Hecla*'s commander would be a dashing young lieutenant, George Lyon, who was used to hardship. He'd barely survived a mission to the desert interior of North Africa in which a companion had died.

Parry was determined to keep his men occupied and entertained. This time trunks of theatrical costumes were packed aboard, along with a printing press, a magic lantern that could project coloured pictures on a bed-sheet screen, and a full supply of library books. Parry decided to establish a school during the long Arctic nights to teach his crew to learn to read their bibles.

He expected to spend at least two winters in the Arctic and he piled provisions for three winters on board his ships just in case he had to spend a third. To help stop scurvy, he proposed to grow great quantities of mustard and cress, and he ordered stacks of hot frames for that purpose. Fresh vegetables were known to ward off the disease.

He would fight off the cold and the dampness with a

newly-designed "Sylvester Stove" that would carry warm air to every part of the hermetically sealed ship. Because there would be lots of fuel this time, it would burn day and night. He also improvised new footwear – using canvas tops and insulated cork soles – and he supplied deerskin jackets for his men.

"Oh, how I long to be among the ice!" exclaimed Parry with all the zest of a schoolboy. His route would be different this time. The only other known avenue leading into the Arctic was the original route taken by Henry Hudson, way back in 1610. It was just possible that a Passage might be found leading westward out of Hudson Bay.

The most likely opening was Repulse Bay, which had repulsed other explorers in the previous century. Still, it had only been partially explored. Was it really a bay? Perhaps like Lancaster Sound it might be a strait. That was Parry's initial goal.

Once again at Deptford the crowds swarmed aboard Parry's ship, *Hecla,* to walk the decks and touch the railings that had once been encased in ice, and to bask in the ordeal. Parry organized a grand ball aboard the sister ship *Fury,* which was decked out for the occasion. As the band played on the upper deck, the company danced on and on into the night under a rising moon, each man and woman convinced he or she was in the presence of adventure.

Ten days later – April 27, 1821 – both vessels were ready to sail. When they reached Hudson Strait, Parry sent a final letter home. He wrote, "I never felt so strongly the vanity, uncertainty, and comparative unimportance of everything this world can give, and the paramount necessity of preparation for another and a better life than this." The Arctic had made him humble.

The only known way to reach Repulse Bay was to circle around the western shores of Southampton Island at the very top of Hudson Bay. Parry, however, decided to gamble by taking a short cut through the mysterious Frozen Strait, which lay to the northeast.

That was unknown territory. Some didn't believe the strait existed. Half the available maps didn't even show it.

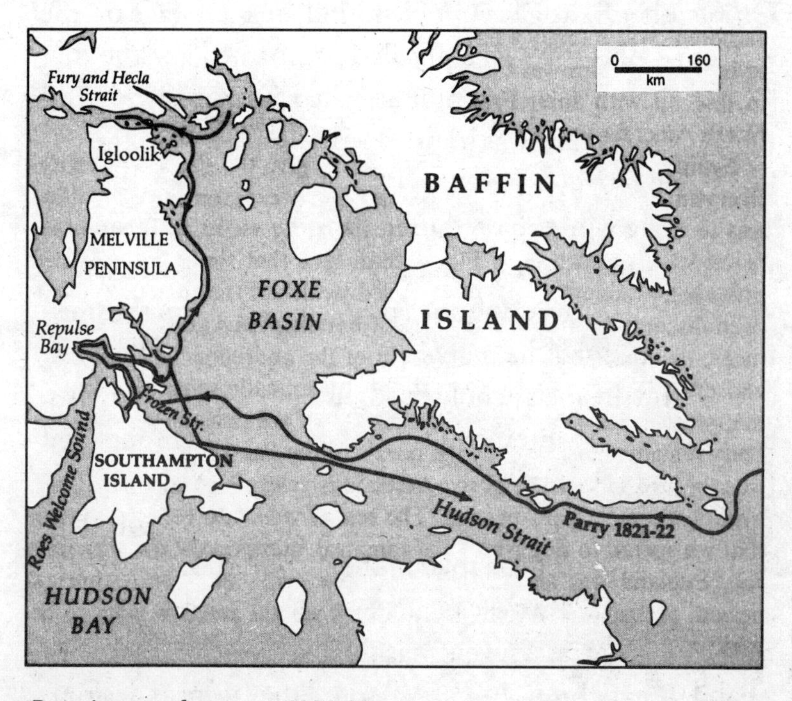

Parry's second voyage, 1821-22

Parry wasn't even sure that he had reached the entrance to it, and so plunged blindly on in a thick fog and fierce blizzard. Then he found to his surprise that he actually got through the strait without knowing it. And it wasn't frozen at all.

He had hoped Repulse Bay would be the link to the Arctic. Now he found that it was land-locked, a dead-end. This was not the route to the Passage. If one was to be found, it must be farther north.

For the next six weeks Parry searched for a way out. But he could find no promising inlet that would lead westward. On October 8, he gave up. He found an anchorage on the east coast of Melville Peninsula (also named for the first lord) and anchored at a point he called Winter Island. And there once again he was imprisoned by the ice until the following July.

The long winter passed more comfortably than the one Parry had endured on the earlier voyage. The new stove worked well. Scurvy was not a problem because of the mustard and cress that he managed to grow. Nor was the crew affected by the melancholy brought on by the long nights, because Parry's wintering place was much farther south and the sun didn't completely vanish.

And there were diversions. The officers shaved off their whiskers to play female roles in the theatre. The school was a great success. By year's end, every man had learned to read. But the greatest event was the arrival on February 1 of a band of sixty Inuit. As Parry wrote, "they were as desirous of pleasing us as we were ready to be pleased." Soon there was singing and dancing on the decks as the newcomers made repeated visits to the ships. The presence of these strange human beings made much of the winter bearable. As Parry noted, the natives "served in no small degree to enliven us at this season."

Still the North West Passage continued to elude him. He sent Lyon off on a fortnight's sledge trip up the Melville Peninsula to seek an opening to the west – a journey that left all the travellers badly frostbitten. Lyon couldn't find the Passage, though he thought there might be a route around the peninsula to the north.

Desperate to get his ships free of ice, Parry kept his crews toiling for three weeks to saw a channel out to open water. Two men died, perhaps from the effects of the work. But on July 2 he was able to set his course north.

He reached a native village, Igloolik, at the top of Foxe Basin. And there he encountered another barrier. Fortunately the Inuit turned out to be expert map-makers. Their simple charts convinced Parry that a Passage existed to the west.

Once again he sent Lyon across the ice with a band of Inuit to pick up fresh fish and to assess the chances of getting through. Lyon enjoyed living with the natives. He learned to eat their food. He danced with the native women and taught them to play leapfrog. He even allowed himself to be tattooed in the native style.

But he found no open water. The ice was still as thick

On the deck of the Hecla *Parry's crew enjoys an Arctic ball.*

as three feet (1 m) with the land obscured by fog. Parry's patience was wearing thin. He was convinced he was at the threshold of the Passage, but he couldn't move.

The Inuit maps had indicated the presence of a narrow fjord. Did that actually lead to the open sea? Parry decided to find out for himself. On August 18, 1822, he stood on the north point of the Melville Peninsula overlooking the narrowest part of the inlet the Inuit had shown. Toward the west where the water widened, he could see no land. He was certain he had discovered the polar sea, and he was convinced that he could force his way by this narrow strait into the west. All he had to do was wait for the ice to clear.

But the ice did not clear. The weather grew warm. An eastern breeze sprang up. But the ice refused to budge. By late September, with a bitter gale blowing in the northwest, Parry gave up.

He was bitterly disappointed. He had waited until the last moment, clinging to the belief that a miracle might occur. But there was no miracle. When he called his officers together they all agreed to remain at Igloolik for another winter and try again the following summer. They could not know that it would be eleven months to the day before they could once more break free of the encircling bonds of ice.

Chapter Four

The people of Igloolik

For the next ten months Parry and his men were in almost daily contact with the two hundred Inuit who lived at Igloolik. Indeed, the most valuable thing about this strangely disappointing expedition were the accounts that Parry and Lyon both brought back of the natives' customs and society. These provided the basis for later studies of the Melville Peninsula Inuit.

Both officers were privileged to observe an aboriginal society untouched by European civilization. It was a society that had managed to exist and even thrive in one of the harshest environments on the globe. Of course, Parry and Lyon weren't anthropologists. Indeed, anthropology – the scientific study of the human race – didn't exist and wouldn't until the mid-century. But both were keen observers. They liked the Inuit. In their long and monotonous confinement they had the time to examine a culture they found foreign and fascinating.

Of course, they judged the Inuit in terms of their own civilized standards. It didn't occur to them – nor would it occur to any Englishman in that age – that different conditions require different codes of conduct.

Parry discovered that the cheerful natives "maintained a degree of harmony among themselves which is scarcely ever disturbed." That being the case, he thought, they could only benefit from Christian evangelism. But he was soon to notice that the Inuit far to the south, who had been "civilized" and Christianized, had turned into thieves, pilferers, and pickpockets, so greedy that one even offered to sell his two children for some trade goods.

By contrast, the uncivilized natives of Southampton Island and the Melville Peninsula were honest to a fault. If you dropped a handkerchief or a glove, they ran after you to return it. Sledges could be left unguarded without fear of loss. Lyon once purposely left a stock of knives, scissors, looking glasses, and other coveted objects in a native hut, and then wandered off, leaving a dozen natives behind. When he returned, he found his possessions intact and carefully covered with a skin.

To both officers the most unusual aspect of the Inuit character was its lack of passion. They weren't war-like or quarrelsome; these typically European emotions were curiously lacking. Feelings of love and jealousy were also apparently unknown. Lyon felt, in fact, that the Inuit didn't possess much of the milk of human kindness. Sympathy, compassion, gratitude – these qualities didn't appear on the surface.

But there was a reason for this. Death was so much a part

of the Inuit life, they had become used to it. In a pitiless land, there was no room for pity. Three days of mourning were allowed after a death and the mourners all cried real tears – but only for a minute. They seemed indifferent to the presence of death. Nobody bothered to cover corpses. The British thought the Inuit callous, but in the Arctic, where exposure, starvation, and disease killed so many so young, no other attitude was possible if sanity was to be maintained.

Parry remarked on what he called the "selfishness of the savage." He thought it one of their greatest failings. The British showered presents on the natives and fed them when they were near starvation, but were annoyed because nobody said thank you.

Obviously it never occurred to any Inuit to acknowledge a gift or a service because in their own world they had to depend on one another. You helped a man out one day, he helped you out the next. That is the way the Arctic world worked. No one was expected to acknowledge kindness.

The natives cheerfully helped the British. They hauled water on sledges, they showed them how to build a snow wall around the *Fury*. They drew maps of the coastline. They brought in fresh fish. They expected presents in return, but to say thank you would have been redundant.

Their own doors were always open. Their food was always shared with strangers – and they didn't expect any payment. They accepted tragedy as they accepted death, with indifference, and sometimes even a little laughter and high spirits. A man could leave his dying wife, not caring who looked after her in his absence. A girl could laugh at the suffering of a dying brother. A sick woman could be blockaded inside a snow hut without anybody bothering to discover when she died. Old people with no dependents were simply left to eke out a living or die. This "brutal insensitivity," as Lyon called it, was appalling to the English, who couldn't comprehend the savage conditions faced by the people of Igloolik.

As they discarded pity, they also discarded the harsher emotions. Revenge was unknown to them, as was war. They didn't quarrel among themselves. They couldn't afford the luxury of high passion. They needed to conserve their feelings to wage the daily battle with the wild. They learned to laugh at trouble, and they laughed and grinned a great deal even when life was hard for them, as it usually was. They lived for the day – for any day might be their last.

Parry thought them wasteful and so they were – in his terms. Life for them was feast or famine. When food was available, they ate it all. When there was none, they went without and didn't complain. The British thought them gluttons. But gluttony in that spare land was one of the few luxuries they knew.

They were always thirsty, and when they could they drank vast quantities of water, or other fluids. For raging thirst was as common in the Arctic as in the desert. To eat snow was taboo for whites and natives alike – the resulting loss of body heat could kill a man. But snow could only rarely be melted because fuel was as precious as food. Water was a luxury to be obtained at great expense.

Parry once conducted an experiment to find out how much a native could eat. He offered a young Inuit, Tooloak, as much food and drink as he could consume overnight. In just twenty-one hours, eight of which were taken up by sleep, Tooloak tucked away ten and a quarter pounds (4.6 kg) of bread and meat and drank almost two gallons (9 L) of liquid.

The irrepressible Lyon decided to pit his man, Kangara, against Tooloak. Kangara managed to devour in nineteen hours just under ten pounds (4.5 kg) of meat, bread and candies, and six quarts (6.8 L) of soup and water. Lyon insisted that if Kangara had been given Tooloak's extra two hours, he would have "beaten him hollow."

To Parry the native diet was "horrible and disgusting." After all, they ate raw blubber. Lyon, who had nibbled sheep's eyes with a Bedouin of the western desert, wasn't so choosy. Like the Inuit, he ate the half-digested contents of the stomachs of caribou, which they called *nerooka.* He found it "acid and rather pungent, resembling as near as I could judge, a mixture of sorrel and radish leaves." But he didn't think to ask why, nor did he seem to connect this half-digested vegetable diet with the natives' remarkable freedom from scurvy.

The natives were just as repelled by British food. They hated sugar. They spat out rum. When one was offered a cup of coffee and a plate of gingerbread, he made a wry face

Tooloak "wins" the eating contest.

and acted as if he was taking medicine. One miserable woman, who had been left to starve after her husband's death, was brought aboard and offered bread, jelly, and biscuit. Lyon noticed she threw the food away, only pretending to eat it.

If the Inuit mystified the British with their customs and attitudes, they in turn were confused and baffled by the strange men aboard the big ships. One thing they couldn't understand was why the strangers hadn't brought their wives with them. When told that some had no wives, they were astonished. Surely, they thought, every man in the world had at least one wife!

They couldn't understand a community whose members were not related. In their own society, *everybody* was related by blood, or adoption. To solve that problem, Lyon told them he was father to the whole crew. That, of course, didn't satisfy some of the women who knew that some of his "sons" seemed older than he.

Nor could they understand the British class system. It was clear that Parry and Lyon were important men. The Inuit believed they owned their ships. But the different ranks confused them. In their society everyone was equal. But, in spite of this clash of cultures, the two peoples got along famously. The Inuit were immensely helpful to Parry and his men, who in turn were generous to them.

If the Britons thought of themselves as the natives' superiors, there is evidence the Inuit thought the opposite. Parry noted that "they certainly looked on us in many respects with profound contempt; maintaining the idea of self sufficiency which has induced them ... to call themselves, by way of distinction *Innuee,* or mankind."

To the British, the Inuit were like children – untutored savages who could only benefit from the white man's ways. This attitude was quite unjustified. In the decades that followed, the real children in the Arctic would be the white explorers. Without the Inuit to care for them, hunt for them, and guide them through that chill, inhospitable land, scores more would have died of starvation, scurvy, exhaustion, or exposure.

Without the Inuit, the journeys to seek out the Pole and the Passage would not have been possible. Yet their contribution has been noted only casually. It was the British Navy's loss that it learned so little from the natives. Had it paid attention, the tragedies that followed might have been averted.

Here was a nation obsessed by science, whose explorers were charged with collecting everything from skins of the Arctic tern to the shells that lay on the beaches. Here were men of intelligence with a mania for figures, charts, and statistics, recording everything from the water temperatures to the magnetic forces that surround the Pole. Yet few thought it necessary to inquire into the reasons why another set of fellow humans could survive, year after year, winter after winter, in an environment that strained and often broke the white man's spirit.

The British felt for the natives. They lamented their

wretched condition. And they couldn't understand why, on being offered a trip to civilization, they flatly refused the proposal. Actually, in most instances the white men were far worse off and much more wretched than the natives. The Inuit were more practically clothed and more efficiently housed. They enjoyed better health than the white explorers, to whom the tough overland expeditions had brought exhaustion and even death.

The Inuit wore loose parkas of fur or sealskin. But the British Navy stuck to the more confining wool, flannel, and broadcloth uniforms with no protective hoods. The Inuit kept their feet warm in sealskin and mukluks. Even Parry rejected Navy leather. The Inuit sleds were light and flexible. The Navy's were heavy and awkward and hauled by men, not dogs. No naval man ever would learn the technique of dog driving or the art of building a snow house on the trail.

Most puzzling of all was the inability of the Europeans to understand the great Arctic scourge – scurvy – that struck almost every white expedition to the North. The seeds of scurvy were already in Parry's men, in spite of lemon juice and marmalade. Yet no one connected the Inuits' diet with the state of their health. Though the effects of vitamins were unknown, the explorers sensed that scurvy was linked to diet, and that fresh meat and vegetables helped ward it off. But nobody caught on to the truth that raw meat and blubber are very effective antidotes against scurvy.

Why this apparent blindness? Part of it was because the

upper classes of England considered themselves superior to most people – whether they were Americans, Hottentots, or Inuit. Part of it was also fear – the fear of going native. The idea of traipsing about in ragged seal furs, eating raw blubber, and living in hovels built of snow did not appeal to the average Englishman. Those who had done such things in some of the world's distant corners had been despised as misfits who had thrown away the standards of civilization to become wild animals.

Besides, it was considered rather like cheating to do things the easy way. The real triumph consisted of pressing forward against all odds, without ever stooping to adopt the native style. The British officers enjoyed these strange child-like people whom they had spent so much time with; but they refused to copy them. And for that they paid a price.

Still, when the Inuit began to leave in the spring in April, 1823, Parry and his officers missed their company and perhaps even envied their nomadic life. The natives were on the move. The white men were imprisoned on their ships, caught fast in the ice.

Parry now decided to send one ship home and carry on alone. But the winter had been appallingly cold. The ice showed no signs of budging. The telltale signs of scurvy – blackened gums, loose teeth, sore joints – were making their appearance. Parry thought that cleanliness and exercise would help stop the disease. In that, he was quite wrong.

August arrived. They were still frozen in. Again, the

crews toiled to saw a channel through the ice pack trying to reach open water. Weakened by illness and by eleven months of being cooped up on the ship, they couldn't work with the same energy. Parry climbed the masthead of his ship and gazed off to the west. And then his heart sank, for, as far as he could see the ice stretched off unbroken. Now he knew that he would have to go home without having achieved his goal.

On August 12, he bade goodbye to Igloolik. He had now been hovering off the mouth of that narrow strait for thirteen long months. He was certain that this was the entrance to the Passage. He was sure the open sea lay less than a hundred miles (160 km) to the west. But he couldn't reach it. The ice-master of the *Hecla* was already dying of scurvy. Others would follow unless he could get back to civilization.

It wasn't easy. Even when he got out of the ice-locked harbour, and fought his way through the pack, there were hold-ups. During one period of twenty-six days, there were only two in which he could move ahead. The scurvy patient died before he could get home.

Finally, on October 10, he anchored off Lerwick in the Shetland Islands, to the ringing of bells and the cheers of the inhabitants, who rushed to the wharfside to greet the ships. These were the first civilized men he and his crew had seen in twenty-seven months. That night the citizens of the little town celebrated by lighting huge barrels of tar on every street.

Alas, Parry's discoveries had been negative. He had learned that there was no route to the Passage by way of Hudson Bay because no ship could squeeze through the ice that clogged the little strait which he named for his two ships – *Fury* and *Hecla*.

Perhaps there were other ways. On the earlier expedition he had ventured briefly down the long fjord he had named for the Prince Regent. Could that be the way? Perhaps at the bottom of the long inlet there was a connection with the mysterious Passage. He would have to mount another expedition to explore that.

His optimism rose. He said he had never felt more sure of ultimate success. He was confident that the English would yet be destined "to succeed in an attempt which has for centuries engaged her attention, and interested the whole civilized world."

In short, he hadn't given up. He was scarcely back home before he was pressing for a third chance to make another voyage. He was obsessed by the mystery of the Passage. Indeed, it obsessed the entire country. No hardship was too unbearable, no years of isolation too stifling, no experience too horrifying to prevent the naval explorers from trying again. One would have thought that a man like Parry might have shrunk from another voyage of dreadful hardships. On the contrary, he was eager to be off.

Chapter Five

The end of the Fury

IN SPITE OF HIS optimism, in the months that followed Parry was at a low point. On his return to England, he learned that his father had died. He was so depressed he couldn't eat or even speak. His sister rushed to his London hotel to find him delirious with high fever. His condition was kept from his mother until the crisis passed.

If that weren't enough, he learned that Miss Browne, with whom he had flirted aboard his ship, and with whom he was said to have an understanding, had lost interest in him. One could hardly blame her. She hadn't seen him since that spring, two and a half years before. But now her mother was going about claiming he had abandoned her.

Parry was miserable. But then he learned that Miss Browne had been seen in the company of other men in his absence. In fact she had actually got engaged to somebody else. And though her mother was trying to get the two together again, Parry wasn't having any of that. In fact, the knowledge of Miss Browne's shocking conduct (shocking to the people of those days only) cured his melancholy.

At the same time he was being praised by the best and brightest in England – all the way from Britain's leading scientist, Sir Humphrey Davy, to Sir Thomas Lawrence, the society painter, and Robert Peel, future prime minister.

He continued to be worried about his future. He really had not been very successful in his quest for the Passage – even though he had gone farther than any other white man. The Navy offered him a minor job, but not as an explorer. But then, in the first week of January, 1824, the decision was made that he sail north again on his quest. Once again, the *Hecla* and the *Fury* would be under his command.

He was determined to be married. And he quickly fell in love with the nearest available candidate, a young woman named Jemima Symes, who was conveniently living at his mother's house in Bath. She was very ill and her condition wasn't helped by the fact that Parry was more anxious to be off to the frozen ocean than he was to be with her. But, for the moment at least, she fitted the role of future partner for which the explorer clearly longed.

At Deptford, the *Hecla* again drew crowds; she was now the most famous ship in the Navy. In the three months before Parry sailed, some three thousand persons signed the visitors' register. They came from all over the British islands, and as far away as Vienna. The well-wishers included Prince Leopold of Saxe-Cobourg, the uncle of the future prince consort, two royal duchesses, and, on the last day, the family of Sir John Stanley of Alderley, whose daughter, Isabella, would one day be Parry's wife.

That April, another future Arctic wife, Jane Griffin, (who would marry Sir John Franklin) met Parry at a dinner party and described him to her journal, as "a fine looking man of commanding appearance, but possessing nothing of the fine gentleman ... his figure is rather slouching, his face full & round, his hair dark & rather curling." To her, he seemed "far from light hearted & exhibits traces of heartfelt & recent suffering, in spite of which he occasionally bursts into hearty laughter & seems to enjoy a joke."

She thought Parry was going back north against his own wishes, complaining to her he had seen nothing of the rest of the world. But this peculiar English reluctance was a mask. He was raring to go, but he didn't want to show it publicly. An Englishman in those days mustn't appear too keen; it just wasn't done.

His orders were clear. He was to sail down Prince Regent Inlet, which he had only partially explored on his first voyage. There he was to look for a channel west which would connect with the coastline which John Franklin, his naval colleague, had explored in 1821. If that could be done, he felt, the Passage was as good as conquered.

Again he reckoned without the changeable Arctic weather. He was faced with another one hundred and fifty miles (241 km) of jostling icebergs in Baffin Bay, all jammed together, imprisoning both his vessels and threatening to crush them like eggshells. He had expected to work his way through in a month, as he had in 1819. It took him more than two.

At last, on September 10, he reached Lancaster Sound again, a month behind schedule. Three days later, only twenty miles (32 km) from the entrance to Prince Regent Inlet, the ships were again caught in the ice. The season was almost over. Now he faced a difficult choice: should he try to make a retreat to England? To Parry that was unthinkable – an admission of defeat.

He determined to push on west as far as possible and try to get to the Passage the following year. But a gale drove him back down the sound and right into Baffin Bay. Then the wind changed and the second gale blew the two ships back again.

He was able to find a wintering place at last on the northwest shore of Baffin Island, in the small bay off Prince Regent Inlet. And for the next ten months this bleak coastline would be his home.

He had never encountered a gloomier landscape. There were no cheerful natives here to while away the dreary hours. No animal was seen. Even the gulls that fluttered around the ship were gone. The white plain was as empty of life as it was of colour.

But of course he had prepared for that. There would be costume parties and grand balls, in which men and officers frolicked together in fancy dress. Discipline was relaxed to a certain extent. For the Grand Venetian Carnival, Parry climbed down the *Hecla*'s side enveloped in a large cloak that he did not throw off until all were assembled on the *Fury*'s deck. To the delight of the company, he stood revealed as an old mariner with a wooden leg, whom his sailors recognized as the man who played the fiddle on the road near Chatham back home.

Parry, the amateur actor, kept up the role, scraping on a fiddle, and crying out, "Give a copper to poor Joe, your Honour, who's lost his timbers in defence of his King and country!"

Not to be outdone, his second-in-command, Lieutenant Henry Hoppner, appeared as a lady of fashion, with a black footman in livery, who was revealed to be Francis Crozier, a midshipman aboard the *Hecla.*

And so they capered to the music of their captain's fiddle – monks in cowls, Turkish dancers, chimney sweeps, ribbon girls, and rag men, Highland warriors, dandies, Jews and infidels, bricklayers and farmers, tropical princesses, and match girls, whirling about in quadrilles, waltzes, and country dances – a bright pinpoint of revelry in the sullen Arctic night.

On July 20,1825, they were freed at last from their winter harbour and set sail for the western shore of the great inlet. Now Parry had felt the real voyage had commenced. They were passing shores that had never been explored. The prospect of a speedy passage seemed bright.

But once more the Arctic blocked them. Hugging the shore of Somerset Island, whose crumbling cliffs sent masses of limestone tumbling onto a mountain of rubble at their base, they ran into a stiff gale. On July 30, it grounded the *Fury* on an exposed and narrow beach. She was scarcely hauled free when both ships were trapped.

To relieve boredom aboard his frozen ships, Parry staged amateur theatricals.

A huge iceberg forced the *Fury* against a mass of grounded ice, threatening to tear her to pieces. She trembled violently. Beams and timbers cracked. A crash like a gunshot was heard and her rudder was half torn away. She began to leak badly, but there was no landing place safe enough to make repairs. All her crew could do was to pray to keep her from drowning again and to work the four pumps in shifts until their hands were raw and bleeding.

For two weeks officers and crew sought to save the ship. They tried to tie her to an iceberg, but the bergs were wasted by weather and the cables snapped. They tried to raise her to examine her battered keel, but a blizzard stopped them. Both ships were in danger of being smashed against the headland. Nothing seemed to work.

The crews had reached the breaking point, so exhausted that some fell into a stupor, unable to understand an order. On August 21, Parry was forced to cast off his own ship, *Hecla,* to save her from being driven aground. The same gale drove the *Fury* onto the beach and blocked her exit with huge bergs. It became obvious she would have to be abandoned.

That caused Parry great pain. Everything about this expedition had been a failure. He hadn't found the Passage – hadn't gotten near it. He had explored no more than a few miles of new land. Now he had lost his ship – the one catastrophe the Navy would find hard to forgive.

He sailed off with the *Hecla* crammed with a double complement of officers and men. He left her sister ship and

Parry and his crew struggle to save the doomed Fury.

most of its stores on the beach – Fury Beach it would be called. Perhaps, he thought, these provisions might help some future expedition.

On October 16, 1825, Parry was home with nothing to show for sixteen months of cold and exertion. At the court martial that followed, all were acquitted and the officers praised and flattered. Parry was relieved, but the fact remained: he'd lost his ship.

Jemima Symes was happy to see him safe. She was still sick, but cheerful. However, the relationship came to an end. But Parry still hungered for a wife. "I have always felt a desire to be attached to somewhere," he wrote a friend. "I have never been easy without it, and with less disposition I will venture to say, than 99 in 100 of my own profession, to vicious propensities, either in this or other ways. I have always contrived to fancy myself in love with some virtuous woman. There is some romance in this, but I have it still in full force within me, and never, till I am married, shall I, I believe, cease to entertain it."

His general loneliness, his hunger for love and marriage, and the knowledge of his failure brought on depression. He suffered from headaches. He took drugs to help him. Now he decided never again to seek the elusive North West Passage. He wanted to be free of complications.

He wrote another polar book – a single volume this time and not nearly as long. It did not meet with the chorus of cheers that had greeted his earlier works. Critics were unenthusiastic, for it was clear that nothing had been accomplished on this last expedition.

Parry had said he would never go north again, that he was through with the Arctic. But those were hasty sentiments, uttered at the end of a long and dispiriting battle with the ice. It began to dawn on him that there was one way to restore his battered reputation, by another daring attempt. If not to seek the fabled Passage, then why not the North Pole itself?

Chapter Six

Treadmill to the Pole

By the spring of 1826 William Edward Parry had fallen in love again. He had reversed his decision to abandon Arctic exploration. The optimistic explorer was sure he could reach the North Pole in a single season. Meanwhile he had asked Sir John Stanley of Alderley for his daughter's hand in marriage.

He had met Isabella Stanley through his friend Edward, Isabella's brother. He was clearly seeking a wife when Isabella, a zesty twenty-four-year-old, fragile looking in the style of the period, but undeniably beautiful, was available. By May he was in love, and sure enough of her agreement to approach her father.

Of course, he didn't use the word "love" to Sir John. It would, perhaps, have been considered crude in those formal days, to admit to something so unrefined as passion. He simply said he was "irresistibly drawn towards ... Isabella by sentiments much warmer than those of common esteem and regard...."

After two agonizing months the Stanleys gave in. The

two were married in October. At the same time Lord Melbourne, the first lord of Admiralty, agreed that Parry could pursue his quest for the North Pole.

For Isabella any separation would be agony, and she knew that a long and agonizing parting was to come. Parry expected to leave England in the *Hecla* in April of 1827 for weeks, months, perhaps even years. He remained the incurable optimist. He was convinced he could reach the Pole by way of Norway in a single season and ignored the experts, who told him how difficult it would be.

The old whaling captain, William Scoresby, now entered the picture. Scoresby wasn't nearly as enthusiastic as Parry was about anybody's ability to reach the Pole in a single season. Scoresby was an extraordinary man – perhaps the most remarkable Arctic expert of his day. A whaler like his father, he had been eighteen years at sea, seven of them as a master. He'd been given his first command at the age of twenty-one and was soon known as the bravest and smartest of the Greenland whalers. But he was more than that. In the winters, when the whaling season ended, he took classes in philosophy and science at Edinburgh. And he was inventive. One of his inventions was a pair of "ice shoes" for walking more easily across the rough ice pack.

He produced a paper on polar ice conditions, and also a monumental work which has been called "one of the most remarkable books in the English language" as well as the "foundation stone of Arctic science."

Now this old-time whaler with the weathered face had

become an ordained minister. He scorned those who believed there was an open polar sea north of the ice pack. The idea that beyond that wall of ice was a warm ocean was one no Greenland whaler could accept. They could all see the wall of ice. Why should it suddenly vanish in a colder climate?

But Scoresby was snubbed by the Royal Navy, which had no use for whalers. This was a snobbish attitude. The men who ran the Navy felt that they were the best people for the job. But Scoresby had far more experience. He had once tried to take part in that first expedition to the Arctic with Parry and Ross. But the Navy had refused to allow that. Now, when he tried to issue some warnings about Arctic travel, Parry and the Navy ignored him.

Scoresby had already made it clear the best mode of travel in the Arctic was by light flexible sledges built on slender wooden frames and covered with waterproof skins – the kind the natives used. These should be pulled by either reindeer or dogs, preferably dogs.

He also warned that any expedition seeking the Pole must set out on the ice when it was frozen hard and was relatively flat in late April or early May. Later on when the weather warmed, huge pools formed and hummocks appeared and slush covered the surface, making travel difficult.

Scoresby had sixty thousand miles (96,500 km) of experience travelling through the ice behind him. He had also gone further north than any other white explorer. But Parry ignored him. He built two heavy boat-sleds, each seventy feet (21.3 m) long, with a twenty-foot (6 m) beam, weighing three-quarters of a ton (680 kg) and equipped with steel runners so that they could be dragged over the ice. He didn't take dogs, and the reindeer he bought were never used.

Nor did he take the whaler's advice about making an early start. He didn't plan to set off across the ice until the first of June. Even that late deadline was missed. Imprisoned in the floes off Spitzbergen well to the north of Norway, he spent ten precious days seeking a safe harbour for the *Hecla* and didn't get away until June 21. Yet he continued to be optimistic. "The main object of our exercise appeared almost within our grasp," he wrote.

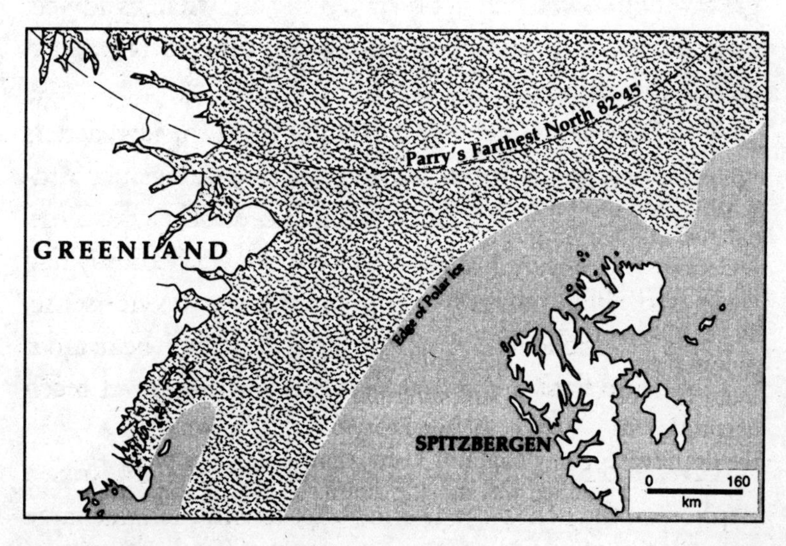

Parry's last voyage: toward the Pole, 1827

He was deeply in love with his new wife. It had been a wrench to part with Isabella and even more wrenching for her, since she found herself pregnant. She couldn't write to him so she kept a diary in which she wrote down her own feelings. She yearned for his presence that spring and spoke to him through her diary: "I would not recall you, your path leads to glory and honour and never would I turn you from that path when I feel and know it as a path you ought to go...."

But the path that Parry had chosen was rough. He expected to find the smooth, flat expanse of ice that some whalers reported. It didn't occur to him, as Scoresby had indicated, that that condition only existed in the early months of the year before the weather changed and the ice grew rougher. Why hadn't he taken the old whaler's advice? Probably because the Navy, snobbish in the extreme, had no use for whalers.

The Navy actually believed that beyond the icy barriers there lay an open polar sea – a ring of warm water surrounding the Pole. That was one reason, no doubt, why Parry built heavy boats rather than the light sledges Scoresby had suggested. The theory of an open polar sea was nonsense. Scoresby didn't believe it for a minute, and when he heard it discussed he flatly predicted that Parry would never reach the Pole.

Parry and his second-in-command, James Clark Ross, a nephew of the discredited John Ross, set off, each in charge of a boat sledge with twelve men. Once again, they were unlucky with the weather. The season was the most unfavourable he could have encountered.

He had never seen such rain. Twenty times as much fell that summer as had fallen in any of the seven previous summers he'd spent in the Arctic. It came down in torrents – once for a steady thirty hours. But then, when the sun came out it shone so hot that the tar ran out of the seams of the boats.

The rain and warm weather turned the ice into a rumpled expanse of broken cakes, all piled one on top of the other, and stretching to the horizon – high, sharp masses that impeded every step of the men. It was like trying to haul a cart through a yard of stones, with the stones ten times their normal dimensions. It was also the "pen knife ice" – needle-like crystals that tore out the soles of boots. And there was slush, knee deep, that caused the men to go down on all fours.

When there wasn't rain there was fog. It was so thick the party couldn't see and had to grope along, yard by yard, from one hummock of ice to the next, trying to avoid the thousands of ponds that formed between the blocks.

The need to launch and land the boats, and to load and unload and reload them, and to make circuits around the ponds slowed the expedition to a snail's pace. Parry had hoped to make thirteen miles (21 km) a day. He could scarcely make half a mile (800 m). On one occasion it took two hours to move a hundred yards (91 m).

He expected to meet what he called "the main ice" – the

smooth continuous plain the whalers had described. He couldn't get it into his head that in summer that didn't exist. His men were wet and exhausted. Their rations were scanty. Parry hadn't brought enough food for the men who were performing hard labour for ten hours.

But there was a worse problem. A stiff wind blowing down from the Pole was driving the ice backwards. At last, Parry understood what the whaling captains always knew – that even as his party plodded grimly north, the ice was moving south. In short, they were on a treadmill. For every step they took forward, the ice took them one step back.

On July 26, Parry realized that, though they seemed to have moved ten or eleven miles (16 to 17.7 km) north, they were actually three miles (4.8 km) *south* of their starting point that day. He didn't tell the men. They thought they were pressing close to the eighty-third parallel at which they would receive a thousand-pound reward that Parliament offered to anyone who got that far.

Two days later Parry had to give up. His instruments told him that he had managed to reach the hitherto unattainable latitude of 82°45′ – a quarter degree farther than anybody else had ever made it. It was a considerable achievement and it would stand for fifty years, adding to Parry's towering reputation as the greatest of the Arctic explorers.

But if he had taken the despised Scoresby's advice he would have achieved more. He and his party had clocked 978 miles (1,574 km) of polar travel, but because of the circuitous route, the need to shovel supplies back and forth, and the southward movement of the ice, he was only 178 miles (286 km) north of the harbour where he had anchored his ship.

And he couldn't go on. His men were suffering from chilblains, snow-blindness, and scurvy. By the time they reached the first solid land, they had been fifty-six hours without rest and couldn't understand orders. Parry noted that they had "a wildness in their looks." They recovered and reached the ship on August 21 after an absence of sixty-one days.

Once again Parry had missed his target. It would have been easier if the Admiralty hadn't felt it necessary to announce he was going to the Pole. "I wish I could say we have been successful but this we have not," he wrote to his wife. He was crazy with desire to "clasp my dear girl to my heart."

As he approached London, he grew more impatient. He landed at Inverness and was held up at Durham for lack of horses. The Duke of Wellington, travelling the same road, had commandeered all available transport. Parry used the delay to express "the unspeakable joy and comfort" his wife's letter had given him. He had received it at Edinburgh the previous night. All the problems of the polar trip vanished when he learned for the first time that he was to become a father.

In a very real way, the birth of his child compensated him for his lack of success. As he wrote Isabella, "success in my enterprise is by no means essential to our joy, tho' it might have added something to it; but we cannot, ought not to have *everything* we wish...."

However, in spite of his failure – or perhaps because he had gone farther north than any other white man – he was knighted and became Sir Edward Parry. He stubbornly insisted that although the Pole would be more difficult to reach than anybody had previously believed, he could not himself "recommend any material improvement in the plan lately adopted." But that flew in the face of all reason and experience.

It was certainly too much for William Scoresby, the whaler, who publicly recorded his disagreement, pointing

out that Inuit invariably used dogs and light sledges and that their light boats were only thirty feet (9 m) long and carried as many passengers as Parry's. Moreover, they could be hoisted on the backs of six or eight men.

The native boat weighed between four and five hundred pounds (181 to 226 kg); Parry's weighed 1,450 (658 kg). It would be many years before the British Navy finally began to adopt Inuit methods of travel in the north, using light sledges and dogs.

This was Parry's last expedition. He settled down now as

The European explorers did it the hard way, with manpower, heavy boats, and poor equipment. The better-dressed Inuit used dogs and kayaks.

the first of the explorer heroes and a member of the famous Arctic Council of senior explorers, which had so much to do with the future exploration of the frozen world. Parry was hailed as the greatest explorer of his day. He had actually gone farther into the Arctic than any other man and farther towards the North Pole than any other white man. Both these records stood for decades.

On that first remarkable expedition luck was with him. In the last his luck failed. But his good fortune and his bad fortune were both tied to vagaries of Arctic weather. He got as far as he did on that first expedition because the weather was on his side. He failed on his second, because the weather was against him.

He was, above all, an optimist. He thought he could go where no man had gone before, and he succeeded. But he also thought he could go farther and in that he failed. When he set out for the Arctic he knew nothing of the Arctic channels or the Arctic conditions. He learned, but he might have learned more had he listened to the natives and observed their techniques.

In that he was very much a product of his time. He was an Englishman and an amateur with all the amateur Englishman's self-confidence and arrogance. It was both his strength and his weakness.

It's ironic that both of the goals that he was seeking on his failed quests turned out to have very little practical meaning. The North West Passage certainly existed. John Franklin came upon it, but died before he could report it.

Robert McClure discovered it and got through it, partly by boat and partly on foot. It wasn't until 1905 that Roald Amundsen went all the way from Lancaster Sound to the Bering Sea – eighty-seven years after Parry's trip to Melville Island.

But the North West Passage had no practical use. Even today with modern icebreakers it's difficult and sometimes impossible to force a ship through that tangle of ice-choked channels. From a commercial point of view it has no value.

And the North Pole? It's simply a pinpoint on the map – a bit of frozen ocean that has, again, no practical value. The nineteenth-century British explorers sought it the same way later explorers once sought to climb Mount Everest – because it was there.

Parry's real contribution, surely, is his careful observations of Inuit life in the days before white civilization changed that way of life forever. His reports and those of his subordinate, Lyon, give us an insight into a remarkable people who, unlike so many white explorers, managed to survive in a dreadful environment. Parry's reputation lies as much in his observations of these cheerful aborigines as it does in his attempts to seek the will-o'-the-wisp of the mysterious Passage. He learned *about* the Inuit; what a shame he didn't learn *from* them.

Index

Also Available

JANE FRANKLIN'S OBSESSION

Little did Lady Jane Franklin know that when her husband, Sir John, left England in 1845 to find the fabled North West Passage, she would never see him – or his crew – again.

By 1847, Lady Franklin was deeply worried about the fate of her husband and his two ships in the Arctic. They had seemingly disappeared without a trace. For the next eleven years, she pressed for a solution to this mystery. Expedition after expedition was sent forth, mapping previously uncharted waters and land masses enroute. When, in 1858, Lady Franklin finally knew the fate of her husband, it shocked the world – and gave birth to a legend.